Get set... GO!

Sally Hewitt

Photography by Peter Millard

Contents

Introduction	2	Wailer	14
Comb and paper	4	Loud and soft	16
Singing balloon	6	Reed	18
Bottles and tubes	8	Buzz and hum	20
Bottle pipes	10	Tongue and lips	22
Pan pipes	12	Index	24

Watts Books

London • New York • Sydney

Introduction

Sound waves are made by
making air vibrate.
This means it moves very fast
to and fro.
You cannot see sound waves,
but you can hear them.

Wind players make air vibrate by blowing.
All the instruments in the picture
have hollow tubes.
Can you spot them?
When you blow into these hollow tubes
the air inside them vibrates and plays a note.

Get ready to make some instruments
to puff and blow.

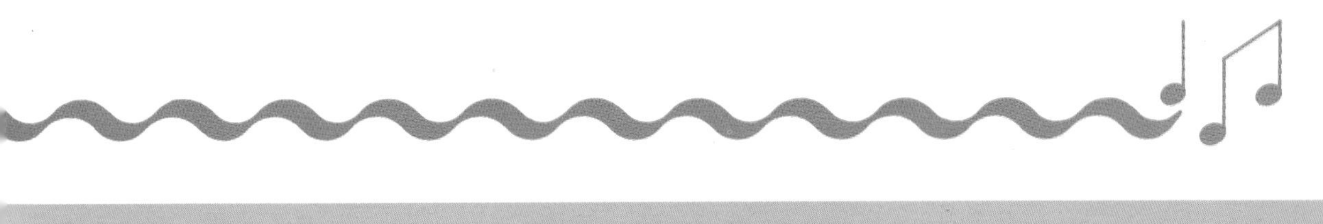

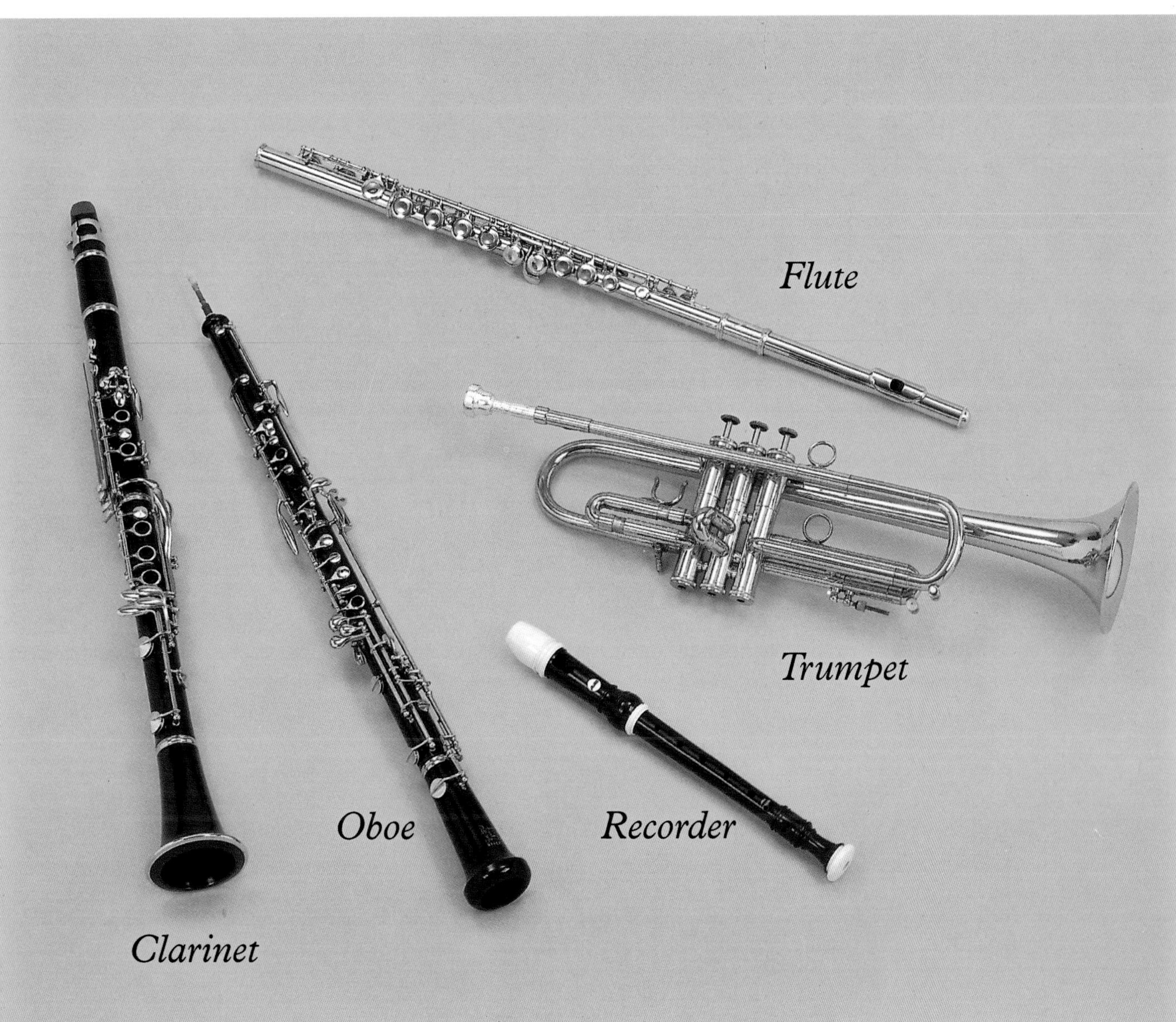
Flute
Trumpet
Clarinet
Oboe
Recorder

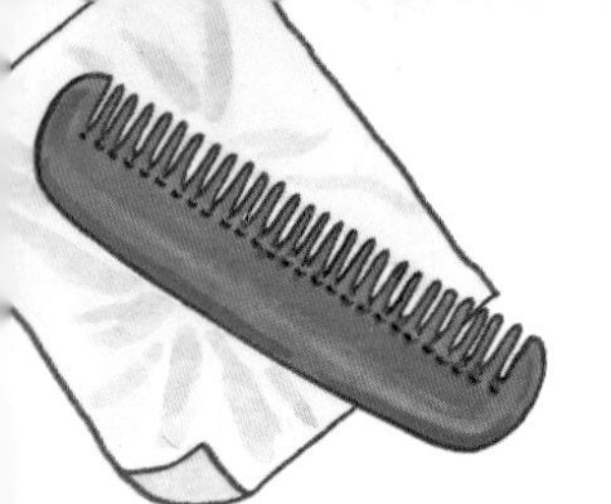

Comb and paper

Get ready

✔Comb

✔Tissue, tracing paper or greaseproof paper

...Get set

Fold the paper over the comb.
Hold them both against your mouth.

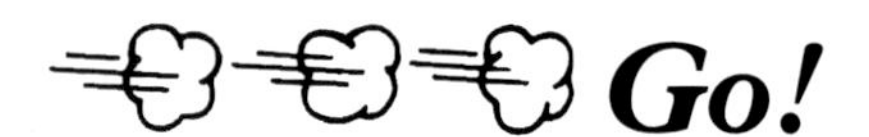

Go!

Sing a tune loudly.
Feel the paper tickle your lips.
Your breath vibrates the paper.
The moving paper changes
the sound of your voice.
Listen to it buzz and hum!

Singing balloon

Get ready

✔Balloon

...Get set

Blow up the balloon.

Go!

Pinch both sides of the balloon's neck.
Let the air out gradually.
The air vibrates as it escapes
and makes a sound.
You can make different sounds.
Let the air out of a small opening.
to make a high squeal.
Let it out of a bigger opening
for a lower wail.

Bottles and tubes

Get ready

✔Bottles of different sizes

✔Tubes of different lengths

...Get set

Sort the bottles by size (big or small).
Sort the tubes by length (long or short).

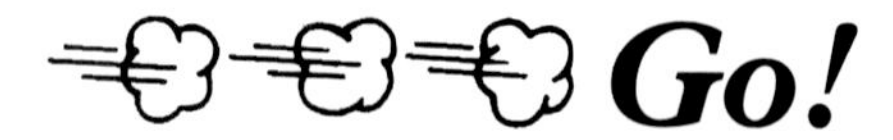

Go!

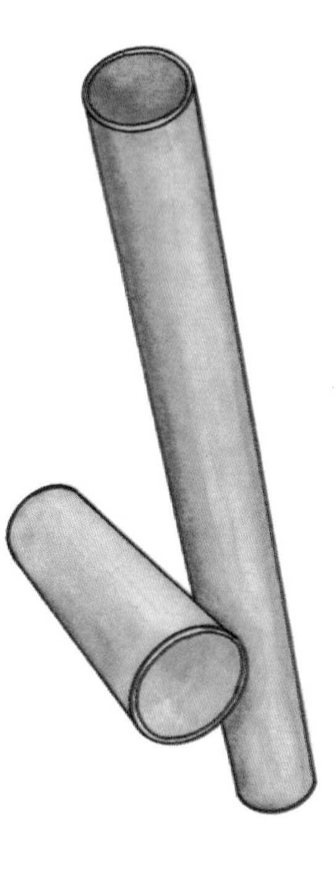

Blow gently across the tops of
the bottles and tubes.
This vibrates the air inside them
and plays a note.
Long tubes and big bottles
make low notes.
Shorter tubes and smaller bottles
make higher notes.

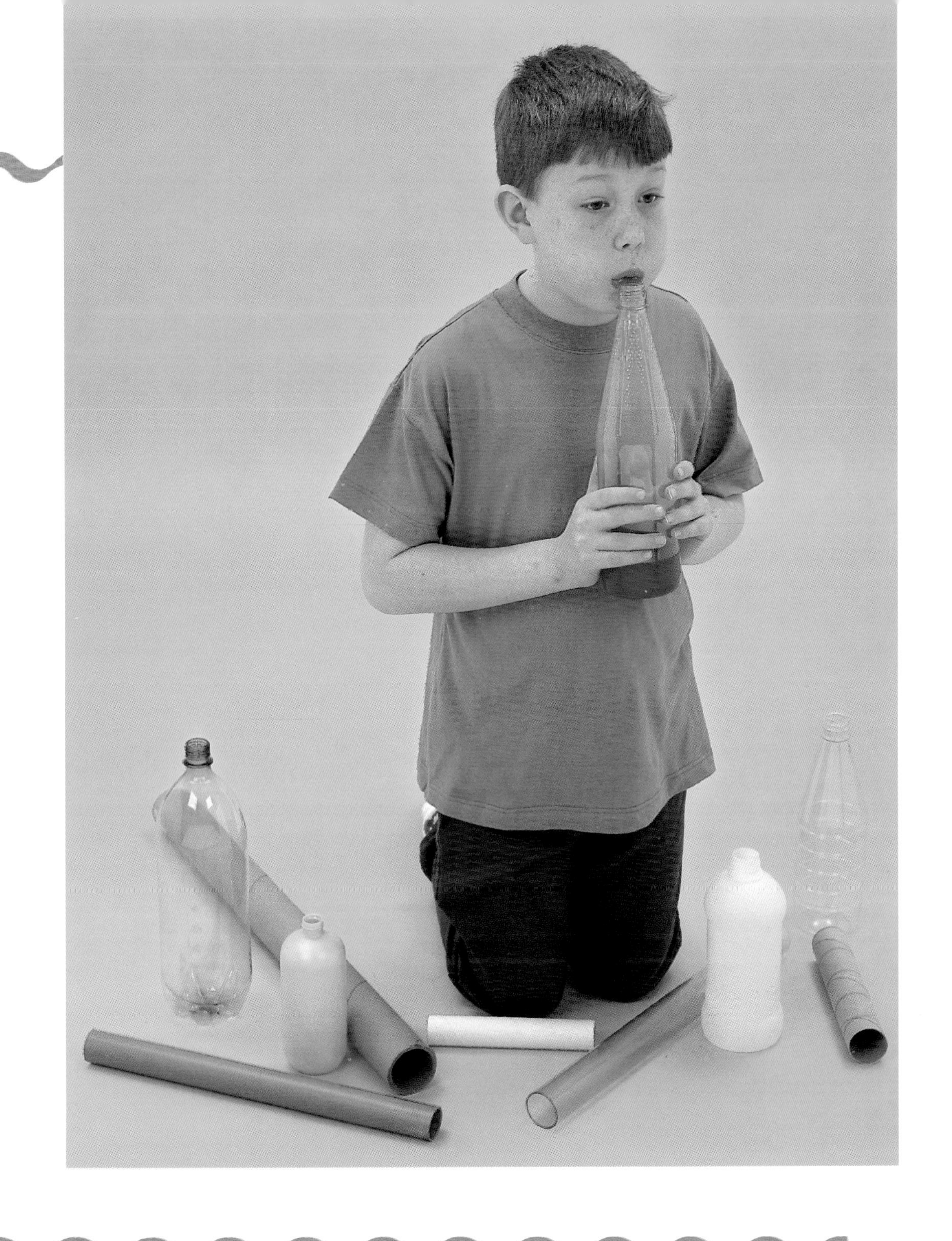

Bottle pipes

Get ready

✔Several bottles of the same size

✔Water

...Get set

Blow across the top of each bottle.
They all make the same hooting sound.

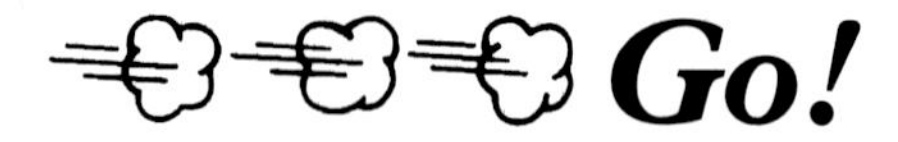

Go!

Pour different amounts of water into the bottles.
Now there is a different amount of air
inside each one.
Blow across them again.
The bottle with the most air
plays the lowest note.
The bottle with the least air
plays the highest note.

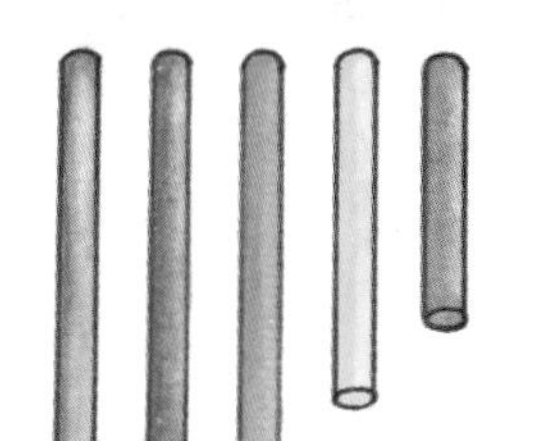

Pan pipes

Get ready

✔Large drinking straws
✔Sticky tape
✔Scissors

...Get set

Blow across the top of the straws.
Listen to the air vibrate inside them and make a sound.

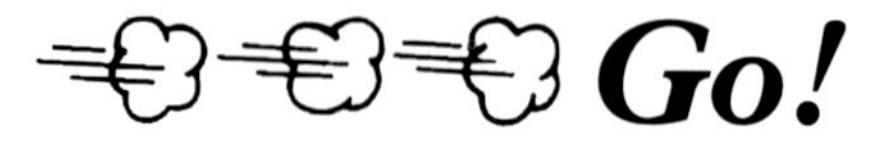

Go!

Cut the straws to different lengths.
Blow each one and listen to its sound.
Short straws play high notes.
Longer straws play lower notes.
Put the straws in order of size.
Tape them together to make pan pipes.

Wailer

Get ready

✔Copper piping ✔Wooden spoon

...Get set

Make sure the handle of the spoon fits snugly inside the piping.

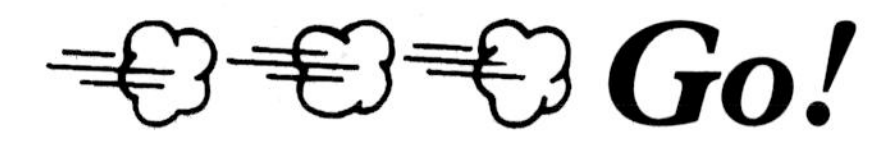

Go!

Blow across the top of the piping.
Push the handle of the spoon
up and down inside the piping
as you blow.
This changes the amount of air
inside the piping.
Listen to the strange wailing sound.

Loud and soft

Get ready

✔Paper tissue

✔Cardboard tube

...Get set

Hold the tissue in front of your mouth.

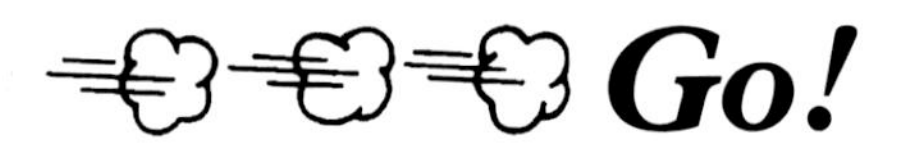

Go!

Sing the sound 'ooo' very loudly.
Watch the tissue move about.
Now sing the sound 'ooo' very softly.
The tissue hardly moves at all.
You use more breath to sing loudly
than to sing softly.
Stuff the tissue into the end of the tube.
Sing loudly down it.
The tissue traps the air inside the tube
and muffles the sound.

Reed

Get ready

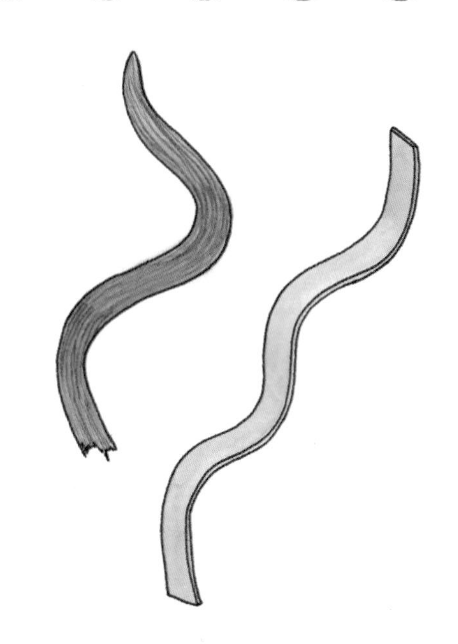

✔ Long blade of grass
(or thin piece of paper)

...Get set

Wash the blade of grass
and dry it thoroughly.

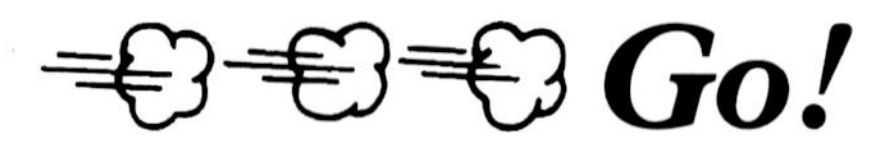

Go!

Hold the blade of grass firmly
between your two thumbs.
Cup your hands.
Press your lips over
the space between your thumbs.
Blow hard.
The grass vibrates and makes
a strange squeaking noise.

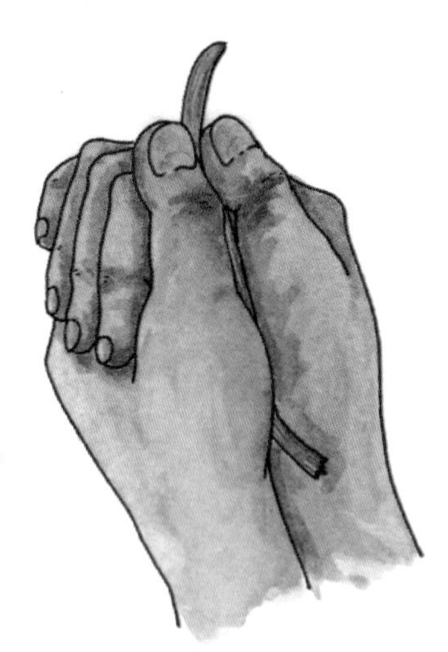

Buzz and hum

Get ready

✔Sing

...Get set

Feel your voice coming from
your chest and head.

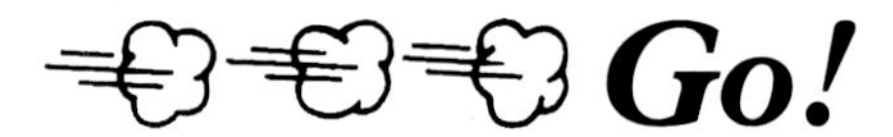

Go!

Make a buzzing sound like a bee.
Put your hand on your chest.
Feel the vibrations.
The sound comes through your mouth.
Now press your lips together
and hum like a top.
Feel the vibrations in your head.
This sound comes through your nose.

Tongue and lips

Get ready

✔Waggle your tongue ✔Move your lips

...Get set

Take a deep breath and sing a long note.

Go!

Change the shape of your lips to make
'ooo', 'aaa' and 'eee' sounds.
Waggle your tongue to make
'lubba-lubba' sounds.
Flap your lips with your finger.
Find out how many different sounds
you can make by moving
your tongue and lips.

Index

air 2, 10, 14, 16
  air vibrations 2
    4, 6, 8, 12

balloon 6
breath 4, 16, 22
bottle 8
bottle pipes 10
buzzing sound 4,
  20

clarinet 3
comb and paper 4

flute 3

greaseproof paper

humming 4, 20

lips 4, 18, 20, 22

note 8
  high note 8, 12
  low note 8, 12

oboe 3

pan pipes 12
piping 14

recorder 3
reed 18

singing 16
sound 16
sound waves 2
sticky tape 12
straws 12

tissue 16
tissue paper 4
tongue 22
tracing paper 4
tube 16
  in wind
    instruments 2

vibrations 20
voice 4

wailer 14
  wailing sound 6,
  14
wind instruments
  2
wooden spoon 14

Watts Books

Books
hard Street
EC2A 4RH

Watts Australia
14 Mars Road
Lane Cove
NSW 2066

UK ISBN 0 7496 1435 8
10 9 8 7 6 5 4 3 2 1

Editor: Pippa Pollard
Design: Ruth Levy
Cover design: Mike Davis
Artwork: Ruth Levy

A CIP catalogue record for this
book is available from the
British Library

Dewey Decimal Classification
620.2

Printed in Malaysia